FURTHER DEVELOPMENTS IN OPTICAL DISC TECHNOLOGY AND APPLICATIONS

R BARRETT

Library and Information Research Report 27

Abstract

This report refers to visits made by the author to selected institutions in the USA in April 1983. The objective of the visits was to take advantage of the presence of the author in the USA on Department of Industry business to update information on current developments in optical disc technology and applications.

Library and Information Research Reports are published by the British Library and distributed by Publications Section, British Library Lending Division, Boston Spa, Wetherby, West Yorkshire LS23 7BQ.

ISBN 0 7123 3038 0

ISSN 0263 1709

The opinions expressed in this report are those of the author and not necessarily those of the British Library.

Typeset by Type Out, London SW16 and printed in Great Britain by DND Business Services, Tadcaster, Yorkshire.
iv

Contents

List of figures

List of tables

Acknowledgements

The visits were made possible by a grant from the British Library Research and Development Department under its Study Visit Overseas scheme. The data presented in the report are based on information given in the form of discussions and technical papers by staff of the following establishments, to whom due acknowledgement is made: BRS Medical, the Library of Congress and the National Library of Medicine. The figures in Chapter 2 are reproduced by kind permission of the Library of Congress.

1 Introduction: optical video discs or digital optical recording discs?

There are two developments of the optical disc referred to in this report, which, although based on the same technology, are distinctly different both in their preparation and application.

1.1 The optical video disc

The video disc development originated from the desire to produce a cheaper and higher-quality alternative to video cassettes for the distribution of copies of full-length movies for playback on normal TV receivers. For this purpose a master video tape of the film is produced which is then used to cut a master disc. The master disc consists of a glass substrate covered with a very thin layer of photoresist material. As the disc rotates, at a speed accurately related to the TV standard, a laser beam modulated by a composite signal formed from the video and two independent audio signals burns holes into tracks on the surface. From the exposed master disc, which has to be prepared in a continuous process, a stamper is produced which is in turn used to produce plastic replicate discs, either in an injection moulding or 'cast and cure' process. The replicate discs are covered with a reflective surface and a thin protective film and two discs are bonded together to form a double-sided video disc.

The same basic process with a different modulation scheme is used to produce compact audio discs.

This process provides a convenient means of distributing relatively cheap copies of video formatted material, but the discs must be produced in an expensive central facility and cannot be altered.

Erasable video discs are being developed and one home-recordable video disc system has been launched, but is not yet readily available, and these systems are not referred to in this report.

1.2 The digital optical recording (DOR) disc

In most digital applications of optical discs, whether for storage of computer data or scanned and digitised images, the requirement is for

1

a system in which the data can be written on the disc as they are generated, stored on that disc for as long as is required, and read back from the same disc at a later date without error. Thus the DOR disc is formatted into tracks and sectors like a magnetic disc, instead of video frames as in the video disc. Also the laser used for writing and reading is a small semiconductor laser, rather than a gas laser as in most current video disc players. The diode laser allows the development of a faster access system but only provides low power. Thus the disc surface must be sufficiently sensitive to allow fast writing and reading with the low-power laser, but at the same time sufficiently robust to allow archival storage of data. The discs are thus much more complex and expensive than optical video discs.

In both the optical video disc and the DOR disc the information pits are fractions of a micron in size, which leads to a relatively high incidence of errors or 'dropouts', due to flaws in the photoresist surface or the presence of dust particles, etc. While this is of little consequence in video replay it can cause large burst error sequences in DOR discs and error correction systems must be used.

1.3 Applications of optical discs

The delay in the appearance on the market of the DOR disc has given impetus to the views put forward by the author and others on the potential for encoding digital information on to the standard TV video signal format in order to produce optical video discs for non-video applications. In addition the interest in interactive video for library applications such as storage of films, photographs and graphics is growing, as is interest in the use of the digital compact audio disc for high-quality archiving of audio production.

The first viable DOR disc system to be made commercially available, though still only in pilot evaluation numbers, is the Thomson-CSF Gigadisc record-read unit, which is to be supplied on an Original Equipment Manufacturer basis for system integrators to build into complete remote access systems.

As Goldstein[1] has pointed out, the DOR disc and the video disc will affect online information retrieval services in a variety of ways, some complementary and some competitive. The DOR disc will be an important adjunct, not only to large centralised services, but also to distributed minicomputer-based systems where purchase of the requisite disc drives can be justified. The video disc, with its potential for publication of machine-readable databases, may provide the greatest competition for centralised online services. Even here,

2

though, the local provision of graphics, audiovisuals and full text may be complementary rather than competitive. The most conspicuous effect of the DOR disc will be in terms of online storage costs. In addition, the very scale of online storage that can be made available will, by itself, generate new offerings and services.

The following reports of activities in the USA highlight developments and progress in these fields.

2 Library of Congress

The Library of Congress has the largest collection of stored knowledge in the world. A significant number of its 80 million items are in an advanced state of deterioration. Most of the printed materials added to the Library's collections every year are on acidic paper; the acid in this paper causes the fibres to become very weak in 25 to 100 years. To preserve items in the original, the Library is developing deacidification technologies and trying to purchase materials printed on alkaline paper.

To provide secondary format preservation and timely public access to high-use material, and fragile or rare materials, the Library has embarked on a pilot programme in image preservation and retrieval using optical disc technology. The three-year Optical Disc Pilot Programme will evaluate the use of optical disc technology for information preservation and management, and determine the costs and benefits of such technology when used in a reproduction setting.

In contrast to microforms, optical disc technology affords extremely high-density online storage of information. A one-sided 30-cm digital disc can store between 10,000 and 20,000 pages of text depending on the resolution required. One side of an analogue disc can store up to 54,000 images. This technology represents a potentially efficient and economical way to store and retrieve images. Optical storage also avoids wear resulting from continuous disc use, since only a beam of light touches the medium during playback. In addition, information on the digital optical disc can be transferred to a new disc without any loss if changes are detected in the original disc. It is thus valuable as a preservation format. Together, these strengths make the disc a potentially ideal medium for on-demand retrieval of high-use items.

The pilot programme was stimulated in part by the Library's use of optical technology in the Cataloguing Distribution Service (CDS), where the images of catalogue cards are captured on an optical disc which can then be used to print exact reproductions on demand. This and suggestions from researchers at the forefront of the technology caused the Deputy Librarian of Congress to initiate several preliminary investigations and to appoint an Optical Disc Storage Technology Committee. This group, using information gleaned from manufacturers, defined equipment and disc specifications and solicited bids. Two contracts resulted: one with Teknekron Controls Inc (now Integrated Automation Inc) to supply an experimental digital disc system, and the other with SONY Video Communications Products

Co to supply experimental video discs and commercial players.

2.1 The DEMAND system of the CDS

The CDS is using optical disc-based technology to store library catalogue card images which are not in machine-readable (MARC) form. There are approximately 5.5 million such cards printed in hundreds of languages and several dozen non-roman alphabets.

CDS has a contract with Xerox Electro-Optical Systems (XEOS) of Pasadena, California for installation of what is named the DEMAND system, a computer-driven optical disc-based mass storage, retrieval and laser printing of non-MARC catalogue cards. The system which was dedicated in August 1982 will gradually replace the mammoth manual system by which CDS has produced non-MARC printed cards for libraries around the world since 1901. The DEMAND system is the second chapter in upgrading CDS's card distribution services with the very latest in electronic storage and printing technologies. In 1978 CDS installed laser demand printing of MARC cards, which makes possible response to orders for any of over 2 million MARC titles within an average of five working days of receipt of order forms. The system uses a Xerox 9700 printer modified to accept card stock. The CARDS system is capable of publishing two pages a second with six cards formatted per page as in Fig. 2.1.

In August 1982, CDS began laser scanning images from non-MARC Library of Congress master cards into the DEMAND system for retrieval and printing on demand in virtual facsimile. This is the first system of its kind based on optical disc storage of highly compressed images using laser input scanning, keyboarding of the Library of Congress card number for control and retrieval of image data, and laser xerographic printing for output processing and distribution at two sheets of six card images each per second. Batches of sheets are guillotined into six separate sequences for processing and distribution.

While the application is oriented to printing, 12 high-resolution cathode ray tube (CRT) displays (1024 lines) are used for editorial and image tagging during the input process. The card image display is very readable despite the fact that only one of every four lines scanned is displayed on the CRT. A very high-resolution CRT would be needed to display fully all the data scanned in and printed out.

Scan in and print out are effectively 189 lines per vertical centimetre of printed display, 71 lines/cm more than the 118 lines/cm resolution that is used in the laser systems for demand printing of MARC cards.

Figure 2.1   On-demand catalogue cards

The higher scan density in the DEMAND system is used in order to capture, for example, the delicate horizontal strokes in Chinese ideographic characters.

Image enhancement is a feature of the input system. The input operator can adjust the scanning laser's sensitivity to grey scale levels in order to prevent stains and other marks on the original image from being stored in the system. In effect, the system reads anything below the grey scale set by the input operator as white space and stores it as such. Obviously a mark on the master image close to the grey scale level of the text itself could not be eliminated without also eliminating the text. In the past, traditional printing technology forced complete type resetting for many cards damaged through the wear and tear of years of use. The image enhancement capability provides significant quality control and markedly facilitates improving master image records.

Since images are being stored as opposed to coded data as in MARC records, it is somewhat difficult to say how many card images can be stored on the optical disc which XEOS is manufacturing and creating. Images are compressed and stored on disc at a reduction of approximately 20 to 1. White space is stored in coded form, not proportionately, thus permitting greater image compression. One estimate projects storing over 200,000 images in digital form on one side of an optical disc. One side of one 36 cm XEOS disc can store 4.7×10^{10} bits, and 5.5 million images may be stored for retrieval on approximately two dozen digital discs. The first full optical disc became available in early 1983 following the successful input and tagging of 200,000 images. Images are stored initially on magnet disc packs (11,000 per pack) to afford correction and interim use in filling orders.

In addition to the file management, retrieval and display activities described above, this technology presents CDS with a powerful system for preserving the master images of 80 years of Library of Congress cataloguing. Optical digital disc storage promises to provide indefinite archival storage for these documents combined with powerful retrieval and facsimile printing. The replacement value of the manual file exceeds $100 million.

2.2   The Optical Disc Pilot Programme (ODPP)

The ODPP has two aspects. Print materials will be stored on DOR discs. Non-print or image-based material will be stored on optical video discs. These two aspects of the programme, involving differing but related technologies are described below.

2.2.1 Print

High-use current periodicals will be given initial emphasis in the digital disc project. Up to 500,000 images per year can be captured, including: materials provided by the Congressional Research Service under its Selective Dissemination of Information (SDI) service (consisting of articles and government documents on public affairs topics); journals in science, technology and business; German, Brazilian, Japanese, Thai, French and Hebraic periodicals; government documents such as the Budget of the United States and the Congressional Record; and United States Agriculture Decisions and Social Security Rulings from 1960 to 1975. Also included will be selected maps, atlases, microforms, manuscripts and sheet music.

Document preparation will be an assembly line process involving keyboarding, scanning and writing to optical discs. At the document preparation station, information relating to the printed library materials will be entered by keyboarding via computer terminals into the Library of Congress Information System (LOCIS) database with a document number corresponding to a Library retrieval number. Documents will then be sent to the input station where they will be scanned and digitised at 118 lines/cm resolution by a high-speed scanner. This digitised information will then be transferred, by a laser writing process, on to the surface of an optical disc. To assure the accuracy of the keyboarding and the precision of the scanning, the information will be written on magnetic disc before being transferred to optical disc.

Because of the high resolution of the digital process, the disc will contain substantially all the information contained in the original item. As a result, the digitised image will be capable of serving as a black-and-white secondary preservation format for the original item, much as microforms are now used. The image will be stored on the disc in much greater detail than it is now possible to display on available high-resolution terminals.

To recall the information, the user will type in the appropriate identification. The Library's computer will tie the request to the image coded on the disc, and display the black-and-white image on the terminal screen. The user stations, to be located in several Library reading rooms, will have a high-resolution terminal for requests and display of full pages of text, along with a medium-speed printer for single-page printout for which there may be a charge. Offline batch printing will be available from a central facility which may also require a fee.

8

2.2.2 Image materials

The analogue video disc is the most promising medium for storing the Library's image materials since up to 54,000 separate black-and-white or colour images may be stored on one side of a 30-cm video disc and selectively viewed a frame at a time. Among the variety of materials in the Library's collections, which are included in this phase of the programme are: 58 large-format colour photoprints that were originally made by adding colour lithographically to black-and-white negatives; five albums of Detroit Publishing Company mounted photoprints of views of the United States and two foreign countries; 600 glass lantern slides; approximately 20,000 glass negatives and transparencies by the Detroit Publishing Company, mostly views and scenes of the United States; 32 paper-print motion pictures; and 91,000 motion picture stills. Several sound recordings will also be transferred to digital compact audio disc.

All the images except for the motion pictures will be filmed on 35 mm motion picture colour film and transferred first to video tape which will then be used to prepare a video disc. The Library will investigate the use of the disc in conjunction with a computer-generated captioning method. In this system, the caption identifying the image would be typed into a terminal connected to a microcomputer and stored on floppy discs. Users would be able to consult the machine-readable index thus created, determine the pictorial items to be viewed and then command the images of the selected items to appear on the screen for viewing. Users would be able to view the image and the caption either simultaneously or separately.

The analogue system, unlike the digital system, will use off-the-shelf technology. It will display items in colour on standard TV monitors. The discs will not replace the original items, but rather act as service copies. Every user does not need to handle the original item; in many cases only the images on the item need be reviewed. In these instances, the discs will allow significant use of the collection without resultant physical deterioration of the original items. The originals will be retained and used only by those with a specific need to study them.

The Library is conscious of its responsibilities under the copyright law. Since this programme raises important copyright issues which will need to be resolved, the Library will be working with the publishing community to be sure that all proprietary interests are protected.

While preservation is initially the primary objective of the ODPP, the discs are also expected to solve the problem of availability of

9

frequently used materials which are not on the shelf. They will also permit viewing of rare and historically important collections not now accessible to the researcher. The Library of Congress is carefully investigating this new technology in order to identify and evaluate different ways of organising, servicing, and accessing important materials in the Library while also ensuring their long-term preservation.

2.3 The current situation

The CARDS system is already in operation, publishing on demand catalogue cards from the MARC machine-readable database, as shown in Fig. 2.1.

The DEMAND system is so far only partially operational. Laser scanning and digitising of the non-MARC master cards is effected at 189 points/cm with a multiple grey level adjustable threshold, and verification is carried out at lower resolution on 1024 line displays. There are currently six terminals inputting data to a PDP 11 computer for temporary storage on magnetic disc and the load on this machine results in very slow throughput. Image enhancement is very good, as was demonstrated by comparison of an enhanced reproduction card with one of a large number of originals which suffered water stains due to a broken water pipe in the storage area.

There has been a delay in the production of the XEOS optical disc system, and in fact the first disc drive had only just arrived at the Library of Congress at the time of the visit and was not operational. Hence copies of the magnetic discs sent to Pasadena for mastering the optical discs are also used to drive the on-demand laser printing facility.

The XEOS system appears to be very much a one-off design, using a very heavy glass disc with a metal centre and enclosed in a protective cartridge. The necessity for mastering in the factory makes it very unattractive compared to the Thomson-CSF Gigadisc system and the Library of Congress is using the latter in the next phase of development.

2.4 Contracts awarded

The Library of Congress has awarded two contracts for the ODPP for information preservation and management. This programme complements the programme of application of deacidification technologies for the treatment of brittle paper caused by acid in the paper-producing process.

10

A plan has been devised, based on a matrix management concept, for managing the ODPP programme in the Library, utilising staff from at least 14 Library offices and divisions to work on 10 project teams, each with a project manager. The Deputy Librarian of Congress will give overall direction to the programme with assistance from a supervisory team. The Chief of the Science and Technology Division will serve as Director of Projects. Four of the project teams, the contract monitoring and design pilot experiment teams, for both the print and non-print projects, have been formed and are currently operating. Approximately 50 Library employees are now involved in the planning and implementation of the programme.

Teknekron Controls Inc (now Integrated Automation Inc) of Berkeley, California has been awarded a contract to provide a system using the Thomson-CSF Gigadisc, which is a microprocessor-based read/write disc drive unit.

Connectable to a Shugart Associates Systems Interface compatible asynchronous interface, Gigadisc gives access to 1 billion bytes held in each face of a self-protected removable disc encapsulated within a cassette.

The Gigadisc drive (Fig. 2.2) consists of:

(a) a solid state laser diode module associated with its optics. A linear motor moves an optical head under the control of radial and vertical servomechanisms. Each position of the head can focus a 40 track wide band into the medium, i.e. a capacity equivalent to 1 megabyte floppy disc, under less than 10 ms access time ($+$ 25 ms average latency time);

(b) a motor and a spindle which maintain the rotation speed of the medium;

(c) a microprocessor logic which controls the SASI interface and, in association with a set of indicators and switches at a front panel, monitors the drive's internal operations.

The medium is a 2×10^9 bytes removable single disc (Fig. 2.3). Information is recorded in a thermosensitive film which is protected within a plastic structure. A track, in spiral, is pregrooved in a substrate and is preformatted with 25 radial sectors. Each sector provides for 1 Kbyte of user's data and its address is prerecorded in the substrate. This structure allows for fast stream-reading/writing operations as well as for the access at random, in read or write modes, to each specific sector.

Figure 2.2   The Gigadisc drive unit

Figure 2.3   The Gigadisc

Writing consists in altering the film by a medium powered laser spot. Reading is achieved by a laser spot generated at a much lower power, therefore at no risk of altering the film, and reflecting from each such alteration.

Because the medium is pregrooved, the same grooving process can be applied to make direct replicates of the discs which hold the user's recorded information.

The system will comprise the disc drive, a multi-disc juke-box, a controller and associated scanners and displays with 118 points/cm resolution. Teknekron gave a demonstration at the Library of Congress in April 1983 of one of the first Gigadisc units linked to a 79 points/cm display (Fig. 2.4). The operating system is due to be installed in spring 1984.

Sony Video Communications Products Co of Lanham, Maryland, has been awarded a contract to deliver 50 copies of each of five analogue optical disc productions of films, video tape, photographs and graphics and 50 digital compact audio discs of two audio productions over the next 18 months.

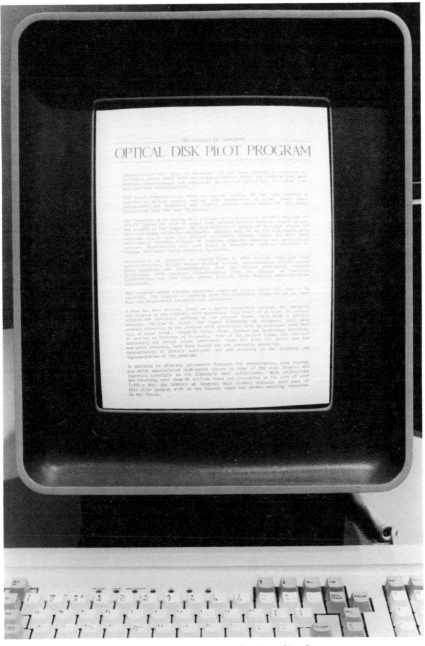

Figure 2.4   High-resolution display

3 National Library of Medicine (NLM)

The Lister Hill Center of the NLM has been involved for some time in a programme of evaluation of optical video discs both for digital applications and as an interactive aid in medical education.

The potential for encoding digital information on the standard TV video signal format has been appreciated for some time[2], allied to the economic advantages of exploiting existing video disc production facilities for the publication of machine-readable information. There are, however, many problems to be resolved before video disc technology can be used successfully for the storage and retrieval of digital information. Paramount among these are error identification and correction. Unlike the DOR process, the mastering and replication process for video disc production does not allow dynamic error identification and correction. Furthermore, there are many known sources of error in the production process, including errors in mastering and stamping and errors due to particulates in the disc plastic. Errors (dropouts) presently occur in the production of video discs for entertainment purposes. These errors, while often visible, are seldom offensive to viewers; however, the sensitivity of digital information to errors is much greater, since the loss or change of a single bit could cause a computer program to malfunction or an index to return incorrect information.

Another major constraint of such recording is the need to encode information into video-formatted frames. This can also be a major advantage, however, since it provides the ability to intermix digital and video information on the same disc. Frames of video, when accessed, will be directed to a video monitor, and frames of digital information will be directed first to the computer and then to whatever output device is desired.

Until recently there has been no evidence of private sector developments in this area. If this technology can be successfully exploited, however, it can serve as a publication process for large databases and full-text document collections, including mixed video and digital information. Toward this end, the Lister Hill National Center for Biomedical Communications, NLM, initiated an R&D programme to address the encoding of digital information on optical video discs and its playback on industrial-type players[2]. The theoretical, error-free storage densities for this process are estimated to be between 20 billion and 30 billion bits, or 2 billion or 3 billion characters per side. The usable data per side will depend on the types and magnitudes of the errors encountered in the mastering and replication of digitally encoded video discs and the error correction methods employed. An experimental facility was established at the Lister Hill Center to facilitate the study of such errors. Investigations seek to determine:

(a) the maximum usable symbol rate per video line;

(b) the maximum usable number of levels (bits) per symbol;

(c) the effect of current video disc production standards on information storage,

that is, the extent to which the theoretically maximum storage is degraded by errors encountered in mastering and replication. Subsequent investigations will seek to identify where improvements in the production process can have the greatest impact on usable information storage.

As part of this R&D programme, the Lister Hill Center developed an intelligent video disc interface unit (VIU) to allow computer control of industrial-type video disc players. Although many investigators are interfacing industrial-type players to microprocessors or computers, the VIU claims to offer improved capabilities, a set of higher-level commands for controlling the player and a degree of device independence. The latter is achieved by allowing different industrial-type players to be interfaced without changing the higher-level control commands used by the computer program.

In actual fact the programme has made little progress, due in part to reductions in funding, and the impression gained is that the NLM is now prepared to await the outcome of developments such as the LaserData system described in the next section.

More progress has been made in the interactive analogue video disc projects described below.

3.1 Videomicroscopy/computer-based learning in medical pathology

A video disc (two sides) containing two single-concept presentations in Basic Medical Pathology became available in 1983. The two programs cover 'Cellular alterations and adaptions' and 'Cell injury'. The discs are recorded in a manner which permits use of the video-micrographs, diagrams, and audio for computer-based instruction (CBI) in pathology. At least one demonstration module for CBI will

17

accompany the video disc. The CBI is constructed to run on microprocessors and there will be two versions of each module: one for single-display device systems and one for systems which use one display device for computer output and a second (TV monitor) for video disc output. If there is sufficient demand, video tape versions of the basic programs will be made available, but there will be no CBI for such taped programs.

The video discs are of the reflective type which will play on several types of playback devices. Any tapes produced will be in the 1.9 cm format. The CBI programs will be written in the BASIC language to run on Apple computers. Single-display-device programs will require a VB-3 Microkeyer (Video Associates Laboratories Inc) to be present in the Apple computer and that requires the computer to have a 64K memory (language card installed). The dual-display device versions of the CBI will run on a standard Apple II with a minimum of 48K of core. Use of the CBI programs will also require a Pioneer (DiscoVision) Model 7820-2 video disc player with a Universal External Interface (UEI) unit.

3.2 Reformatting of the American College of Radiology (ACR) learning file

The ACR Learning File (ACRLF) consists of some 5,000 radiographs covering 1,400 cases and is currently in use at 219 health science teaching institutions. The cost for copies of the file is approximately $35,000.

The ACRLF has been developed over a period of years as a collaborative venture between the Bureau of Radiological Health of the Food and Drug Administration, the American College of Radiology and the University of California at San Francisco. Although it is generally conceded that this collection of radiographs is a highly useful learning tool, its use has been limited by the high cost of producing copies of the file and by the cost of using and maintaining such a large number of radiologic films. Experiments have therefore been designed to determine whether reformatting of the file as video images recorded on an optical disc could produce a version of the file which would be as useful as the current version and less expensive to acquire and maintain. Although previous attempts to obtain acceptable video images of radiographs were not entirely successful, it appears initially that the present effort has produced images which will be acceptable to radiologists for teaching, and perhaps for diagnostic, purposes. It is believed that the explanation for this apparent success may lie in the fact that the video-recording was supervised by a radiologist who previewed each image on a video monitor and coached the video equipment operators to adjust and re-adjust cameras and signal processing equipment until an optimum image was obtained.

3.2.1 Reformatting procedures

Video formatting of radiographs from the ACRLF was carried out in the TV production facility of the National Medical Audiovisual Center at the NLM. The radiographs were mounted on standard radiology viewing boxes and matted to shield the image seen by the video cameras from extraneous light. RCA TK-45 cameras were used to produce video images which were recorded on Ampex 5 cm 'quad' and 2.5 cm helical scan video tape recorders.

The zoom feature of the cameras was used in some cases to enlarge significant areas of the radiographs. Whether recording the entire radiograph or some sector of it, in all cases a radiologist approved each picture, as displayed on a monitor, before it was recorded. Multiple frames (15-50) of each picture were recorded.

The original video tapes were edited on to 2.5 cm helical scan tape format and sent to optical video disc manufacturing facilities. The first video disc was mastered by DiscoVision Associates (now Pioneer USA) and the second by the 3-M Company. In the first experiment 15 full-size radiographic images were recorded along with four subsector enlargements. The second video disc has images from seven ACRLF cases and five cases from the Armed Forces Institute of Pathology (AFIP). There are 30 reproductions of original radiographs and 20 sector enlargements in this series. The second disc also contains split-screen images in which multiple radiographs can be seen simultaneously.

3.2.2 Computer control

Optical video discs provide a very large storage capacity for visual information. Each side of a disc has a theoretical capacity of 54,000 pictures. Obviously such storage would be useless without an index to provide access to individual frames. Such access can be achieved manually using a hard-copy index and a manual disc player controller. (A frame number is entered using the key pad, followed by activation of the search key.) A search can also be conducted using an external controller such as a computer. In either case, access time for any single frame is less than 2 s.

In the present project, two types of access control programs have been developed for use with microcomputers (Apple II). First, discs can be searched for individual radiographs listed by case number and radiograph description in an index displayed on the computer screen. They can also be searched by means of programs designed as clinical case simulations.

Since each frame on a video disc is associated with a five-digit number (00001 through 54000), the effectiveness of search and retrieval is essentially limited only by the creativity of the designers of the computer programs.

3.2.3 Practical considerations

In its current format the ACRLF consists of copies of radiographs stored in paper jackets. A printed index organised on the basis of organ systems is associated with the Learning File. The outer jacket for each case has a summary of the case history, laboratory results, and physical findings (where appropriate) types on the jacket. A carefully prepared teaching discussion is also typed on the jacket (upside-down with respect to the history section).

It is intended that a student or resident will study the file by:

(a) selecting a case from the file index;

(b) reading the history and the physical and laboratory findings;

(c) examining the radiographs;

(d) making an interpretation;

(e) checking his/her findings against those of the experts.

Finally, the user is supposed to return the films to the jackets and reinsert them into the file for the next user, but this does not always occur. As presently formatted, the Learning File requires about 0.72 m^3 of storage space plus associated study area which must include several view boxes. As indicated previously, the current cost of a single copy of the file is $35,000. Other parameters associated with the current file are similar to those of any collection of radiographic film (handling ease, half-life, storage conditions such as temperature, humidity and silver content).

If video radiographs similar to those on the discs described here are found acceptable to radiology educators, it would then be possible to proceed to reformat the entire ACRLF on video disc. The cost of such an undertaking would be about $250,000. The cost of a copy of the video disc (*all* the collection could be accommodated on one side of a single disc) would be a few hundred dollars at the most, depending upon whether recovery of the development cost would be required.

The user of such a video radiological learning file would require, as a minimum, the following: a copy of the video disc, a video disc player ($2400), a television monitor, and a hard-copy index to the video disc ($100 or less). For optimum utilisation of the video collection, the user would need the disc, a player, a television monitor, and a micro-computer with interfacing devices ($1500-$2000), and computer soft-ware (on a 'floppy' diskette) to control the disc player ($100 or less).

Obviously, if the picture quality of the video subset of the Learning File is acceptable to radiology educators and the entire file is reformatted to optical disc, considerable cost savings can be realised in the future. In addition to reduced costs, there are other potential advantages inherent in the optical disc format. So far as is known, an optical video disc when handled with reasonable care has a useful life of at least 10 years and probably much longer. It is resistant to rather large changes in temperature and humidity; it is impermeable to liquids and it is not affected by ultraviolet or infrared light.

3.2.4 Educational consideration

In addition to cost and convenience, educational advantages of some magnitude may be expected from use of an optical disc version of the ACRLF.

The individual radiographs on the video disc could be used in the same manner as the ACRLF films have been used in the past. Specific cases could be called up on the basis of organ systems, using a workbook containing the organ system index and printed copies of the histories and expert discussions found on the protective jackets in the present collection.

Alternatively, an external computer could be programmed to control the entire process. The organ system index could be supplied by the computer and once a case was selected by the user, the history and laboratory/physical findings could be called up and displayed on the computer screen. The individual radiographs could then be displayed under computer control in any sequence, as often as desired, and for whatever time period needed by the user. The expert's summary of the case could be displayed by the computer on demand or the user could be required to make an interpretation prior to seeing the experts' findings.

The video radiographic visual database can also be used to construct highly interactive case simulations.

Another possible use for such a visual database would be to construct didactic presentations in which the computer would output descriptions of radiologic problems, using the video radiographs much as slides are used in lecture/conference presentations.

Any complete video radiology disc would contain, in addition to the cases in the ACRLF, a complete set of normal films which could be called up at will by any user and by case simulations or didactic presentations programmed for external computer control.

Finally, such a visual database could be used to construct examinations for either self or formal evaluation purposes.

4 BRS Medical

BRS, with its headquarters in Latham, New York, is a division of Indian Head Inc, a billion dollar conglomerate based in New York City, which was recently purchased by the international firm Thyssen Bornemiza. BRS's efforts are concentrated in:

(a) managing a public search service offering international access to databases;

(b) designing, developing and maintaining databases for the private use of individuals and organisations;

(c) designing and implementing sophisticated, specialised information storage and retrieval software.

Since 1977, the BRS Online Search Service has provided high-quality retrieval programmes, special pricing options and unique participation in service management. BRS currently develops and maintains a wide variety of databases for government agencies, private companies, academic and research organisations throughout the world.

BRS Medical was launched in 1981 to answer the information management needs of the medical profession. Its first development for clinical practitioners is the Medical Information Retrieval Service, which became available early in 1983 under the name COLLEAGUE and links full-text searching of medical publications by computer with display of illustrations, charts, X-rays, etc, from a video disc. This is similar to the Pergamon VideoPatsearch system developed for patent searching and, in fact, BRS Medical did a lot of the development for Pergamon.

MIRS has been developed to give medical practitioners facilities for searching and retrieving information in clinical settings and not just in libraries, where most database search facilities are located. Fast access is available to a comprehensive library of medical knowledge, including Medline and Pre-Med, and ER/ICU with the complete text of major medical reference books and journals, rather than just abstracts. ER/ICU also includes lists of concise data on differential diagnoses, signs and symptoms, drug interactions and side effects, and diagnostic protocols.

4.1 LaserData

In a recent report[3] the author referred to research being carried out in the Architectural Machine Group of MIT into possible digital applications of the analogue video disc. BRS Medical became aware of these developments, set up a company called LaserData and hired the graduate research team from MIT with the objective of turning the standard optical video disc player into a microcomputer peripheral, providing digital data storage and retrieval facilities from a commercially produced analogue disc, termed DataDisc. By devising a means to encode high-density digital information within the video signal on the analogue video disc and using standard disc mastering and pressing facilities, it is claimed that comparative costs for the system are achieved as in Tables 4.1 and 4.2

Table 4.1   Storage media, capacities and costs

Medium	Characters of text (millions)	Cost per million characters ($)
Removable Magnetic Disc Pack (CDC9766)	256.0	4.00
Floppy Disc (20 cm double sided, double density)	2.5	2.00
Magnetic Tape (high density, 2460 b/cm)	150.0	0.20
Microfiche (CDM, 48X)	1.6	0.09
LaserData DataDisc (L-4800)	4,800.0	0.02

Table 4.2   Memory/controller hardware, capacities and costs

Medium	Characters of text (millions)	Hardware cost per million characters ($)
Magnetic Disc (IBM 3380, Winchester type)	1,200	35.00
Optical Memory Drive (STC,DRAW-type disc)	4,800	25.00
LaserData 100 Controller (includes DataDisc player)	4,800	1.30

The primary market LaserData is addressing is the publishing industry, specifically publishers of reference works, directories, texts and periodicals for medical, legal, scientific and other professionals. A variety of applications are possible since the DataDisc is claimed to accommodate full text (searchable), combined text/illustrations and video in both still-frame and movie format.

LaserData sees itself as having two distinct roles. The first is to provide the service which encodes the information on to the DataDiscs. The second is as an original-equipment supplier of a selection of hardware — largely for resale by publishers and information distributors to their customers. BRS has exclusive licence to all LaserData technology in biomedicine.

4.1.1 Products offered

The LaserData line of products includes nine distinct models tailored to the differing professional needs of potential end-users. Most are configured in 'work station' form and employ a selection of optical video disc players controlled by a professional-standard microcomputer. Product offerings include:

LaserData 100 Controller—the necessary hardware to enable a microcomputer and optical video disc player to become a knowledge retrieval system of extraordinary power and flexibility.

LaserData 101 Professional—a complete work station encompassing a microcomputer with associated hardware, one terminal and DataDisc player. Designed for use by small medical, legal, engineering or other professional concerns with a continuing demand for database information but modest resources.

LaserData 102 Researcher—an expansion of the capabilities of the 101 Professional, this work station includes four DataDisc players to quadruple the online capacity for data search and retrieval. Configured to meet the expanded needs of reference libraries, corporate researchers and other users of large databases, it includes one terminal and can be expanded to serve four simultaneous users.

LaserData 103 Networker—a further refinement of the 102 Researcher to enable it to function as a 'file-server' for a local-area network. Designed for use in either microcomputer networks or a mainframe environment.

LaserData 201 Professional Plus—an upgrade of the 101 Professional to include a separate video display monitor for access to both video and mixed digital-and-video output.

LaserData 202 Research Plus—a similar upgrade, applied to the 102 Researcher.

LaserData 203 Instructor Plus—designed for use in tutorial and sales-training applications, it includes a microcomputer terminal with associated hardware, four DataDisc players and a separate video display monitor with 'touch-sensitive' screen for interactive programs.

LaserData 301 ImageMaster—a complete digital image retrieval and analysis work station including image processing software, high-resolution display and four DataDisc players. Configured for medical scanning imagery, remote sensing/cartographic data, and other digital image applications.

4.1.2 Services provided

The *DataDisc Mastering Service (DDMS)* is a proprietary means of encoding clients' information onto a LaserData DataDisc and then duplicating that master DataDisc in the quantities specified by clients.

LaserData DataDiscs are available in four data capacities, as shown in Table 4.3

Table 4.3 DataDisc capacities

LaserData DataDisc	Characters of text (millions)
A-800	800
L-1600	1,600
A-2400	2,400
L-4800	4,800

The *DDMS* includes a selection of discounts for both quantity orders and repeat business from clients.

Additionally, a number of optional services are available under the aegis of *DDMS*.

The *DataDefender* encryption process ensures complete control of access to data, no matter how widely it is distributed. Separate 'keys' can be assigned to different databases contained on a single DataDisc so that customer access can be sold database by database.

The *DataMax* data-compression techniques allow a further increase in the text storage capacity of the DataDiscs by an additional factor of more than 2, with no loss of integrity.

The *ExComp* process allows creation of marketable databases directly from past and present computer photocomposition tapes, thereby providing new opportunities to sell in database form information which was originally acquired for conventional publishing purposes.

The *DataMark* service ensures that a database, if not already in searchable form, can be searched 'To the word' by customers in seconds on a microcomputer.

The *DataSource* service enables conversion of past and present hard-copy text into machine-readable text files in DataDisc form.

The *ImageEnhancing* service includes registration, resampling and other value-improving processing to enhance image usefulness, facilitate overlay techniques and minimise the work station processing load once a customer has selected an image to be viewed.

The *DigitMaster* digitising service can build image databases from hard-copy sources, both by digital scanning and point-by-point on digitising tablet.

4.2   The current situation

As far as can be assessed, the MIRS system is operational, but the LaserData concept is at a much earlier stage. Some $900,000 research and development costs have gone into LaserData, including the taking on of the MIT graduate research staff and it is expected to continue at about $250,000 per annum. LaserData took the DiscoVision system with the objective of achieving some 1 Gbyte of data storage out of a possible 2.2 Gbytes, taking account of error correction requirements due to the poor error performance of the discs. The error protection process is currently a cumbersome but workable process. A VAX 22/70 type computer is necessary to obtain real/time data speed for the video format, but a relatively simple decoder can be used under control of a UNIX type 16-bit microcomputer.

BRS was expecting to market test a completed device including NLM backfiles in selected libraries at the end of 1983. It is also looking at possible use of the compact digital audio disc which should provide some 800 Mbytes of usable data.

The impression gained from BRS Medical was that NLM is likely to cease its experimental work on digital applications of video discs and await the outcome of the LaserData developments. However, no evidence has yet been seen of the success of these developments, or of the actual availability of services and products referred to above.

5 Conclusions

There is no doubt that applications of both optical video discs and DOR discs are gradually beginning to make their mark in the library and information services fields. The significant programme being embarked on at the Library of Congress, relating both to preservation and improved online access, is an indication of the importance of these potential applications in libraries. The importance lies in the possibility of answers to such questions as:

(a) What are the practical limits of the equipment?

(b) How does the average library user react to doing research from a CRT screen?

(c) How important is it to have single-page copying, total-document copying or computer-output microfiche available to the researcher?

(d) What is the optimum design configuration for a user station?

(e) What manpower requirements are necessary to do document preparation and scanning of information into the digital storage system?

(f) What materials are best suited for this type of system?

(g) How will quality be measured?

(h) What replication and life tests will be conducted?

(i) How will new interaction and satisfaction be measured?

(j) What instructional material and assistance will be needed?

(k) How will legal issues, such as copying, be resolved?

(l) What personnel, materials, and facilities will be used?

(m) What procedures will be used for input of material, quality control of scanned images and exception handling?

(n) Is the digital optical disc a suitable long-term preservation medium for printed information?

(o) Is the analogue video disc a suitable long-term preservation medium for image and audio information?

(p) What are the best approaches for storing optical discs to achieve maximum lifetime?

(q) Is it feasible to measure changes in information quality on discs in real-time?

(r) Is it feasible to detect an upper soft error limit on discs which will trigger copying the information on to a replacement disc?

(s) Is it feasible to transfer analogue information to another medium with minimum loss of information?

The LaserData development is also significant and supports the author's views of the important potential for the analogue video disc in the data distribution field. However, it is not yet possible to comment on the success or otherwise of the venture.

6 References

1. Goldstein, C M. Optical disc technology and information. *Science,* vol 215, February 1982, pp 862-868.

2. Barrett, R. *Developments in optical disc technology and the implications for information storage and retrieval,* British Library R&D Report 5623, 1981.

3. Barrett, R. *Optical video disc technology and applications: recent developments in the USA.* British Library LIR Report 7, 1982.

Other reports

Library and Information Research (LIR) Reports may be purchased from Publications Section, British Library Lending Division, Boston Spa, Wetherby, West Yorkshire LS23 7BQ, UK. Details of some other LIR Reports are given below.

Computer software: supplying it and finding it by W Tagg and R Templeton
LIR Report 10   ISBN 0 7123 3014 3
Following the 1981 seminar on 'Libraries and computer materials', a representative survey of the various agencies involved in the production and distribution of computer software was carried out. The results of the survey are used: (a) to give a broad indication of the state of publishing in this area; and (b) to show the various data elements employed in the description and recording of information about software. An examination of the elements in (b) provides the basis for some observations about cataloguing requirements, and for comparison with existing guidelines.

Recent initiatives in communication in the humanities by M Katzen and S M Howley
LIR Report 11   ISBN 0 7123 3016 X
This report comprises a number of contributions concerned with recent activities in humanities communication in the UK and the USA. It includes an account of an Anglo-American conference held on the topic in May 1982. It also covers some initiatives which took place subsequently — in particular, a visit to the USA by Dr Katzen, the establishment in the UK of the Office for Humanities Communication and a proposal to the American Council of Learned Societies for the establishment of an Office of Scholarly Communication and Technology in the USA. In addition, it contains a select bibliography.

An investigation of the use of systems programs in library applications of microcomputers by A Trevelyan and M Rowat
LIR Report 12   ISBN 0 7123 3017 8
For various areas of library activity readily available software tools might be used as alternatives to the writing of dedicated applications programs for systems based on microcomputers. This report examines six areas in detail: circulation control, acquisitions, cataloguing, notification, small databases, and periodicals circulation. The approach adopted offers the basis of a fast and relatively cheap method of implementing systems for particular applications in libraries.

Libraries bring Prestel to the public: a summary of British Library-supported research 1979-1981 by Judy Redfearn
LIR Report 13   ISBN 0 7123 3019 4
Between 1979 and 1981 the British Library supported research into Prestel to see if it could assist in the two main roles of libraries: the dissemination of published information and the publication of local information. Prestel sets were installed in different types of library throughout the country, and public libraries and local authorities were invited to become members of an umbrella group of information providers. This report summarises the aims, discusses the chief findings and assesses the broader implications of the research.

Curriculum change for the nineties: a report of the Curriculum Development Project on library and information work by E P Dudley et al
LIR Report 14   ISBN 0 7123 3018 6
Part one places the report in its historical context and describes the work which took place. Part two contains the conclusions and recommendations relating to the variety of courses, the content of courses and the constraints on change. The report seeks to promote an informed discussion rather than to prescribe solutions.

The schools information retrieval (SIR) project by M E Rowbottom et al
LIR Report 15   ISBN 0 7123 3020 8
The SIR project was designed to introduce the principles and methods of computerised information retrieval into secondary schools. A microcomputer software package was commissioned and then tested in a variety of secondary schools. The report describes the software and the experiences of each of the schools.

Microcomputer applications in academic libraries by Paul F Burton
LIR Report 16   ISBN 0 7123 3021 6
The report describes the results of a survey of microcomputer use in UK academic libraries carried out in 1982. Details of the hardware and software of many of these libraries are included in the report, as are current and imminent applications. There is also an annotated bibliography on microcomputers and libraries.

The economics of information by John Martyn and A D J Flowerdew
LIR Report 17   ISBN 0 7123 3022 4
A group of economists, accountants and information specialists met in June 1982 to discuss problems in the area of economics of information. This report contains the discussion document circulated to

participants before the meeting, the opening address and the final report summarising the discussion.

Libraries and trade unionists: a report on needs and provision in public libraries and elsewhere by Alan Clinton
LIR Report 18   ISBN 0 7123 3027 5
At the centre of this report are the details of two questionnaires enquiring into need and provision for information among trade unionists. The author is critical of the view that trade unionists' information requirements merit only limited consideration, and draws attention to examples of good practice in meeting trade unions' needs. The report also makes suggestions about acquisition, organisation and publication.

Local area networks: the implications for library and information science by Mel Collier
LIR Report 19   ISBN 0 7123 3028 3
Local area networks are systems allowing high-speed interconnection of computers within a restricted area. They facilitate distributed processing and indicate a trend away from large, centralised computers. Microcomputer networks are being offered as alternatives to minicomputers for substantial data-processing activities. This report, commissioned in 1982, gives definitions, describes concepts and introduces some of the techniques involved. The role of local area networks in library and information science is examined, current initiatives are reviewed and finally some suggestions are made for further research in the field.

Inventory of abstracting and indexing services produced in the UK by J Stephens
LIR Report 21   ISBN 0 7123 3030 5
This inventory updates BL R&D Report 5420 of the same title. The entries are arranged alphabetically by name of service. Four indexes are provided: broad subject headings, specific subject headings, an index of responsible authorities, and an index of database processors with the UK online databases they offer.

Information demand and supply in British industry 1977-1983 by The Technical Change Centre
LIR Report 23   ISBN 0 7123 3033 X
Between November 1982 and April 1983 a study was conducted into the effects of the current recession on the supply of technical and commercial information services in British industry. The purpose was to see how industrial information services have adjusted and to

investigate how information providers outside industry have reacted to any changes in demand from industry. Questionnaires were distributed among 238 library and information service units in industry, and among 305 external information providers of various kinds. Most than half of these provided responses that were analysed. There is evidence that the pattern of demand for, and supply of, information in British industry has changed substantially, and possibly permanently, since 1977.

Preservation policies and conservation in British libraries: report of the Cambridge University Library Conservation Project by F W Ratcliffe with the assistance of D Patterson

LIR Report 25   ISBN 0 7123 3035 6

The Cambridge Conservation Project had two immediate objectives: to establish the facts about preservation policies and practices in libraries in the UK and to identify the educational and training facilities available to librarians and practitioners. Nationwide surveys by questionnaires, interviews and seminars were among the methods used. The report makes recommendations for action in two areas, first within individual libraries, involving little or no additional expenditure and immediately applicable, secondly at a national level. Among the latter, the twin needs for cooperative action, for which no mechanism exists at present, and for a focal point for preservation, some sort of National Advisory and Research Centre, are of prime importance. The status, funding and location of the latter need further clarification but involvement of the British Library in any such undertaking seems essential to its success.